WILDLIFE!
WISHES!

ARKANSAS WILDLIFE
Intimate Portraits of Species That Roam "The Natural State"

Tim Ernst

Cloudland.net Publishing
Cave Mountain, Arkansas
www.Cloudland.net

Goldfinch on a grapevine (previous page)

Coyote eyes

Manufactured in China
Library of Congress Control Number: 2009906550
ISBN: 9781882906666

Book designed by Tim and Pam Ernst
Species identification coordinated by Don Kurz

American bald eagles (facing page)

INTRODUCTION

Welcome to my presentation of ARKANSAS WILDLIFE! We have an amazing collection of wild critters that call "The Natural State" home, including nearly 700 vertebrate species (mammals, birds, reptiles, amphibians, and fish), plus thousands of invertebrates (insects and worms). I offer 124 photographs of 80 different species here for your viewing pleasure.

The majority of these images were produced specifically for this book project and were made within the past couple of years by me (my lovely bride took the photograph on the table of contents page). I have included a handful of favorites as well that I made "accidentally" during my 34-year career as a nature photographer. There are a few hints in the text about photography, plus a longer discussion near the back of the book. I am primarily a landscape photographer, but tuned my equipment and my lifestyle towards wildlife for this book project.

The purpose of this book is to feature portraits, sometimes very personal and intimate, of the beautiful wildlife species that can be found in Arkansas. It is not a technical guidebook, scientific journal, or meant to document individual animals in their native habitat. While most of the photographs are of wild animals in Arkansas running free, not all are. A small handful were captive animals, pets, or in one case an injured bird that was being rehabilitated (none were zoo animals though!).

The common name is listed with each photograph, plus one or more facts that I found interesting or educational about each species. Sometimes I got a little carried away since there was so much information about some of these critters—you might want to first sit down and look through just the pictures, then when you have more time go back through and read the text. I have not listed specific location information unless it was important.

It is my hope that through this presentation you will gain an increased sense of appreciation for the great wildlife that we are blessed to share our state with. "The Natural State" is not just about the land and the water, but also about the wondrous critters who live here as well. It is up to us to protect this precious resource for future generations, and I hope that you will take the opportunity to visit some of our wildlife preserves and parks and see the wildlife in person.

E*njoy!*

Tim Ernst at Cloudland
July, 2009

Tim Ernst

Red fox and her pup

CONTENTS

Spicebush swallowtail butterfly on sunflower (photo by Pam Ernst)

Snow geese, Mt. Nebo, and the setting full moon
Holla Bend National Wildlife Refuge

Wildlife can be found just about everywhere in Arkansas, including in your own back yard. Habitat destruction due to the progress of man has taken its toll on many species though, and without our help much of the wildlife we see and enjoy today would have become extinct or only found in zoos.

We have one of the best collections of wildlife refuges of any state in the country, including almost 100 wildlife management areas run by the Arkansas Game and Fish Commission, ten national wildlife refuges run by the U.S. Fish and Wildlife Service, plus more than 50 state parks and millions of acres of national forest land where wildlife can flourish.

Some of these areas were specifically set up for and do protect wildlife (no hunting is allowed in most of the state parks), but the simple truth is that most of our wildlife lands exist for the production and management of game—that means hunting and fishing. Most of our wildlife was pretty much wiped out by the early settlers who did not realize that wildlife had to be managed in order to survive. There is no question that without hunters and fishermen paying the bills for nearly 100 years we would have very little in the way of wildlife in Arkansas.

We have made great strides in recent years with non-game species too, thanks in part to a special 1/8th cent conservation sales tax that we all pay. Wildlife agencies spend more and more time and money managing habitat and protecting species than ever before. Besides having a lot more wildlife around to enjoy, we also have wildlife areas that are equipped for wildlife watching with trails and blinds, and several large nature centers have been built around the state where you can learn a great deal about the native wildlife species.

Many of the photographs in this book were taken at these wildlife areas, including this one of geese taking off at the Holla Bend refuge near Russellville, one of my favorite haunts. I encourage you to visit these areas every chance you get!

Black bear

Bears can sprint up to 35mph in the short run, although sometimes they have to stop and scratch an itch!

Hickory horned devil—regal moth caterpillar

This guy is among the largest of our native caterpillars (five to six inches in length—about the size of a large hot dog). He has five pairs of horn-like structures on his head area and short spikes on the rest of its body. When disturbed, these caterpillars throw the front part of their body from side to side in a ferocious display, using the spikes to scare away predators, and people too! Quite harmless though.

American white pelicans (whose bills can hold more than their bellies-can)

In some Asian countries fishermen will tie the throats of pelicans shut, then train them to fly off and fill their bills with fish and return to the boat fully loaded.

Wild turkey

Turkeys are great runners, often choosing to flee danger on the ground instead of taking to the air. In fact they can even outrun a fox. Benjamin Franklin wrote a famous letter in 1784 criticizing the choice of the bald eagle as our national symbol, and suggesting the wild turkey as a better representative of American qualities. He wrote the turkey is "a much more respectable bird, a little vain and silly but a Bird of Courage." They taste pretty good too!

Five-lined skink

The eggs have a thin, elastic, parchment-like shell that is easily punctured. During the incubation period the eggs absorb moisture and swell up considerably as you see here. Note about feeding a pet skink—feeder insects should be "gut-loaded" before being offered to the skinks: chicken mash is good for this purpose. I think I'll stick with my springer spaniel, Aspen—he is easy to feed! Juveniles of this and other skink species have bright blue tails (that eventually fade) and are often referred to as "blue-tailed" skinks.

White-breasted nuthatch

We call these guys "upside-down" birds since we often see them working down a tree trunk. They get their common name from their habit of jamming large nuts and acorns into tree bark, then whacking them with their sharp bill to "hatch" out the seed from the inside. They are monogamous and mate for life.

White-tailed deer fawn

Fawns are born with no scent and are taught to drop and curl up when danger approaches. Most predators don't see color and the fawn's pattern hidden away in tall grass or forest floor keeps them from being seen. This photo was one of my first wildlife images, and the fawn was "hiding" only a few feet off of the hiking trail I was on. He never moved, but I'm sure mom was keeping a watchful eye nearby.

Immature American bald eagle

The famous white head and tail feathers don't arrive until an eagle reaches sexual maturity during the fourth or fifth year of life. Until then a young eagle will maintain a striking pattern of contrasting browns and blacks with whites and grays—still a stunning sight to behold!

Crayfish pincher

This old warrior probably lived a good long life before being eaten by a raccoon. Their presence in a stream is a good indication of water quality as they cannot tolerate polluted waters. We often associate them with Cajun cuisine, but crayfish are a popular dish in Scandinavia, too. The boil is typically flavored with salt, sugar, ale, and large quantities of the flowers of the dill plant. While most Americans eat them warm, the Swedish and Finnish normally eat them cold.

Common raccoon

Raccoons often take their food to a nearby stream or pond to wash it carefully before eating it (although they are also known to "wash" their food in *dirt*, so the washing may not be a sign of actually cleaning the item). They often find snacks in the water too, like crayfish! For climbing down a tree headfirst, a raccoon rotates its hind feet so that they are pointing backwards.

Monarch butterfly on purple coneflower

The monarch life cycle is one of the most amazing stories in all of nature. It takes several generations in a single year to sustain the species, and each generation goes through four different stages of life—egg, caterpillar, cocoon, and finally the adult butterfly that we all know and love. The adults that pass through Arkansas in the fall begin their migration up north, not stopping until they reach their winter home in Central Mexico. In the spring the butterflies head back north, stopping along the way to lay eggs on a milkweed plant as they pass through Arkansas, which begins the next generation. The adults of the next couple generations will only live a month or two, moving farther north with each cycle. The last generation that is born in the fall up north is the one that will make the long trip south to spend the winter, beginning the process all over again.

Mallard duckling

Ducklings go through a long incubation period in the egg and are more mature than other birds and can swim and feed themselves as soon as they hatch.

Trumpeter swans at Magness Lake

No one had seen a wild trumpeter swan in Arkansas in more than 75 years. Then on December 30, 1990, three were spotted on a small lake near Heber Springs. Each year the swans have come back to spend the winter months on the lake, bringing more buddies with them. The latest count was 134 swans in 2009. They arrive at the 30-acre Magness Lake each year within a day or two of the full moon in November, and begin heading back to their nesting grounds in Minnesota and points north in February. This lake is privately owned, but access is granted to the public to view the swans, and daily feeding continues (of the swans, not the tourists!).

In December, 2005, I photographed two adult trumpeter swans at the Mill Pond in Boxley Valley—the second location in modern recorded history for the swans in Arkansas (see them on page 87). They returned the following winter with a baby (called a cygnet), much to the delight of everyone passing through Boxley. They returned for a third winter in 2007, however the adult male was killed at the pond by an unknown critter and the remaining family has never returned.

During the winter of 2008, banded juvenile swans from Iowa were released at the Mill Pond in Boxley and at Holla Bend National Wildlife Refuge near Russellville. These "zoo" swans as I call them never did fly back north the way that normal wild swans do, but rather remained in Arkansas the entire year (and again in 2009). The hope is that they will fly north and then bring wild swans back to Arkansas with them. Perhaps they will succeed next year!

These truly are amazing and spectacular birds and if you ever get the chance to visit them you should do so. Directions to all three locations are included in the ***Arkansas Nature Lover's Guidebook***.

Female purple finch (previous pages)

This small bird uses its big beak and tongue to crush seeds and extract the nut, and it is fun to watch them do this at a backyard feeder. This particular girl is shown here right outside my office window coping with the terrible ice storm of 2009—we kept the feeders filled!

Mayfly hatch

The lifespan of an adult mayfly can vary from just 30 minutes to just a day or two depending on the species. It often happens that all the mayflies in a population mature at once (the hatch) and the mayflies will be everywhere, dancing around each other in large groups, or resting on every available surface. Then poof, they are all gone!

Woodchuck

How much wood could a woodchuck chuck if a woodchuck could chuck wood? It turns out the answer is about 700 pounds.

Black bear, cinnamon variant

About a third of the black bears in Arkansas are actually cinnamon or brown colored their entire lives.

Summer tanager

This is the only bird in the United States that is completely red. They spend their winters in Panama or South America. The summer tanager is considered a bee and wasp specialist. They usually catch a bee in flight and then kill it by beating it against a branch. This particular bird was my alarm clock for several months. His beautiful song began at the break of day, and continued for a couple hours.

Gray tree frog

This is one of the many tree frogs that lives near our cabin. When we have a serious thunderstorm roll through, our group of frogs will tell us when the storm is about to end—they are quiet during the worst part of the storm and heavy rains but will strike up the band as the storm is moving out, letting us know it is safe to plug our computers back in again!

Prothonotary warbler

This bird was named after officials in the Roman Catholic Church known as the protonotarii, who wore bright yellow robes. They winter in South America and make the long journey to Arkansas in the spring to nest. These tiny birds live high up in swamps and are hardly ever seen. Then all of a sudden one will drop out of nowhere and say HI! A beautiful jewel indeed and one of my favorite birds of all.

Spotted ladybug and toothwort wildflower

There are over 5,000 species of ladybugs worldwide. Certain species lay extra infertile eggs with the fertile eggs, which provide a backup food source for the larvae when they hatch. A ladybug beats its wings 85 times a second when it flies. Ladybugs make a chemical that smells and tastes terrible so that birds and other predators won't eat them. And finally, just in case you needed to know, a gallon jar will hold about 80,000 ladybugs!

Red-tailed hawk

These hawks often soar with wings held high, flapping as little as possible to conserve energy. If the wind is right they can hover and remain stationary above the ground, watching for any movement on the ground. When the time comes, they change the shape of their wings and can attain a speed of 120 mph while diving after prey. Many hawks live around our cabin and it is such a joy to watch them soar and play—sometimes I think they know we are watching and put on a spectacular show just for us!

Rocky Mountain bull elk and his harem

Sometimes wonderful things happen right out of the blue, and that is serendipity. Such was the case one crisp fall day as I was wading up the Buffalo River in search of beautiful fall color. I was standing knee-deep in the middle of the river photographing a neat reflection when I noticed some movement out of the corner of my eye—a herd of cow elk were approaching the river. As luck would have it I was using a wide-angle lens and my telephoto lens was in my camera bag, across the river on the bank. The elk would have been tiny specs with the wide-angle lens so I frantically waded across the river to get my big lens, but that put me out of sight of the elk. So I waded back out into the middle of the river once again and then discovered there was a bull with the cows! I quickly set up my tripod and big lens, and just then the bull turned and looked directly at me (he was probably wondering who this fool was standing in the river with expensive camera equipment!). I only had time to take *one* photograph before the bull turned away and walked off.

Green tree frog

These guys love to sing in the rain, and always seem to have a friendly greeting when you get to see one up close. And yes, the most famous frog of all is a green tree frog—Kermit!

Funnel web spider

These spiders construct "sheet" webs that are shaped like a funnel, then they will hide in the hole and wait for some action. When prey hits the web and gets stuck the spider rushes out and grabs the prey, dragging it back into the funnel hole for dinner. We had hundreds of these funnel webs appear literally overnight around the cabin this spring—and were filled with dew drops as you can see. They disappeared the next day, not to be seen again for the rest of the season—very odd.

Canada geese at sunrise in the delta

These geese mate for life with very low "divorce rates," and pairs remain together throughout the year. They love to fly at sunrise!

Long-horned bee on purple coneflower

Many of the popular wildlife species are the big ones like deer, elk, bear, bald eagles, etc. But we also have many thousands of different species of small wildlife too that is every bit as colorful, interesting, and active.

Take this little bee for example. It's easy to see where he got his name from—look at those long horns! Can you see his eyes? And look closely at his legs—they are covered with tiny hair that serves a great purpose on the world stage. This bee spends his time crawling all over flowers like this one, and as he does, tiny pollen particles cling to those little hairs—you can see them on his legs and on his butt too. When this bee visits another flower he will deposit that pollen as he crawls around, thus helping to pollinate the flowers!

To get this photograph I stood in the hot sun for an hour holding a large camera and long macro lens with twin-flash units attached to the end of the lens. In order to get this close I had to lean into the flowers just inches away—there were bees buzzing all around me, sometimes even crawling on my pant legs! These bees were in constant motion and it was tough to keep up with them and get a photo that was in focus, but I had a blast and really enjoyed watching the bees go about the business of life.

There are tiny wildlife species everywhere—next time you are out in the woods, or in the back yard, drop to your hands and knees and crawl around a little bit and see what you can find!

Elk herd in a pasture of meadow buttercups (previous pages)

There is one dominate bull with this herd, plus another bull that will eventually challenge the big guy for the favor of the ladies.

Rose-breasted grosbeak, male
The male grosbeak will share the incubation duties and sit on the nest during the day, taking turns with the female.

Rose-breasted grosbeak, female
Both birds sing quietly to each other while they switch places on the nest.

Acorn weevil grub

An adult weevil will drill a tiny hole into a live acorn that is still attached to a tree to feed on the meat inside. After eating their fill, females often lay eggs in the acorns on which they've fed. Legless, grub-like acorn weevil larvae hatch from the eggs a few days later. There may be one to several acorn weevil larvae in each acorn. Larvae typically go through five growth stages. Each stage ends with the shedding of the old skin, providing the larva with more room to grow. After a few weeks the larvae chew their way out of the acorn, sometimes right into the waiting arms of a flying squirrel!

Southern flying squirrel

This is the most common squirrel in the Ozarks, but we seldom ever see them since they only come out at night. The term "flying" is somewhat misleading, since flying squirrels are actually gliding mammals incapable of sustained flight. They will jump from a high branch and glide to the next tree, sometimes going as far as 60 feet. They have a furry membrane which extends between the front and rear legs, which helps them glide through the air. Steering is accomplished by adjusting tautness of the membrane. The tail acts as a stabilizer in flight, much like the tail of a kite, and as an adjunct airfoil when "braking" prior to landing on a tree trunk. This particular squirrel above, named Pete, belongs to some friends of ours who raise flying squirrels. He is very old for a flying squirrel (11 years and counting), and has enjoyed many fine meals of captured acorn weevil grubs that we've brought him over the years. In the wild these squirrels will chew out a smooth round hole in an acorn trying to find the little grub—that's a telltale sign that a flying squirrel is nearby!

Lessor yellowlegs

These guys are active feeders, often running through the shallow water to chase their prey. It is great fun to just sit and watch.

Black bear with twin cubs

Bears don't hibernate in Arkansas—they go into a deep sleep and spend the winter months inside their dens, which is when cubs are born—what a great deal for mom—go to sleep in the fall and wake up in the spring with babies! Twins are common, and we've even seen a bear with triplets near our cabin before. Four cubs are rare. Bear dens are most often dug into the hillside in rocky areas, but bears will also den in large hollow trees.

Baby river otter

When otters submerge themselves their eyes are shielded with a clear eyelid. This special trait allows the otter to see clearly underwater, while the eyes are protected. They can dive to a depth of 60 feet.

Pileated woodpecker

This is the largest woodpecker in the United States and it is quite startling to see one in the deep woods since they are *so* big! They dig rectangular holes in trees to find ants, and these excavations can be so broad and deep that they can cause small trees to break in half. This bird is often reported to wildlife officials as being the infamous Ivory-billed woodpecker (sorry, no photo). But their real claim to fame is that they were the model for one of my favorite Disney characters that will live on forever, Woody Woodpecker.

Mallard duck

This is our most common and recognizable duck. They are highly gregarious outside of the breeding season and will form large flocks with thousands of birds. The duck hunting season has become one of the greatest social events in Arkansas, and we are considered the duck hunting capitol of the world. (Ducks love rice, and we produce a *lot* of rice!)

Horsefly and beaked moss

Ouch! Sorry to be a tad bit brutal here, but I don't like horseflies and tend to mash them up whenever I get the chance (merely killing one doesn't always disable them). Unlike insects which puncture the skin with needle-like organs, horseflies have mandibles like tiny serrated, curved swords, which they use to rip and/or slice flesh apart. This causes the blood to seep out as the horsefly licks it up. They may even carve a chunk completely out of the victim, to be digested at leisure. This is the only critter that will keep me from going on a hike if I know they are about—they can bite through clothing and love the taste of insect repellent!

Yellow garden spider

I don't normally like spiders, but often will stop and inspect any in the orb weaver family like this fat and juicy guy—they are always so colorful! This spider builds a zigzag band of silk through the center of their web (seen in the lower part of this photo). It is thought to be a lure for prey, a marker to warn birds away from the web, and a camouflage for the spider when it sits in the center of the web.

Barn owl

The barn owl is one of the most widely distributed birds in the world, found on all continents except Antarctica, and on many oceanic islands as well. Compared to other owls of similar size, the barn owl has a much higher metabolic rate, requiring more food. Pound for pound barn owls consume more rodents than possibly any other creature—a nesting pair and their young can eat more than 1,000 rodents in a single year. Barn owls have a very short life expectancy of only a year or two.

Black bear cub

Momma bears will often use nearby trees as nurseries and send her cubs up the tree for safe keeping while she attends to other business on the ground. I was driving backroads one day and spotted a large momma bear and her cub grazing in the grass along the shoulder of the road. I naturally came to a screeching halt and grabbed my special "wildlife" camera setup and started to take pictures. The cub immediately ran off into the woods, but soon came back out again into the open. Momma did not seem to pay much attention to the cub, or to me. Eventually the mom got agitated with the cub for some reason and chased it off into the woods, where the cub climbed the first big pine tree he could find. What luck—I would be getting a photograph of a cub in the tree!

As the cub began to scamper up the tree (they can climb really *fast*) I noticed some movement above him, and soon a second cub appeared, this one climbing down the pine tree. A moment later a *third* cub appeared—mom had triplets and had sent all three cubs up the tree while she was feeding.

My lovely bride wanted so bad to take this little guy home, but of course, the cute cubs grow up and soon there would not be room at the dinner table!

Striped skunk family (previous pages)

This is one of the most feared critters in the entire wild kingdom! An adult will carry just enough of the nasty chemical we're all so afraid of for five or six uses, and requires some ten days to produce another supply. Skunks love to eat bumblebees. And the young "kits" enjoy playing about as much as anyone.

Egret nest with eggs in flooded rookery

The nest is a small untidy platform of sticks in a tree or shrub constructed by both parents, usually over water. Sticks are collected by the male and arranged by the female, and stick-stealing by other pairs seems to be a form of recreation. This particular colony numbered more than 1,000 nests—each of the white spots you see in the background in this photo is an egret sitting on a nest. I was standing in waist-deep water when I took this photo—I was kind of nervous since there were four large alligators circling nearby waiting for lunch!

Cattle egret in breeding plumage

This is a relatively new species to Arkansas but is expanding rapidly. They like to hang around cattle and eat ticks and flies off of them so ranchers don't mind having them around. Males are typically nondescript white most of the year, but they can get all dolled up like this guy with colorful plumage to attract a mate when the breeding season is on.

Rocky Mountain elk calf
A baby from the larger members of the deer family is called a "calf" and has spots when it is young.

White-tailed deer fawn
A baby from the smaller members of the deer family is called a "fawn" and has spots when it is young too!

Prothonotary warbler

This tiny bird played a role in American history. Back in the 1940's during a House Un-American Activities Committee hearing a young congressman used a mutual sighting by two defendants of this warbler as proof the two men knew each other—and as a result, Alger Hiss (a suspected spy) was later convicted of perjury, and Richard Nixon arrived on the public stage. You could say that the prothonotary warbler helped propel Nixon into the presidency!

Young raccoon

Raccoon "kits" or "cubs" love to play in and around hollow logs. Their characteristic face mask that looks like that of a bandit has enhanced their reputation for mischief—hum, if the mask fits, wear it!

Eastern tiger swallowtail butterfly and red clover

This is such a bright and beautiful butterfly and one of my favorites! The first drawing of a swallowtail in the New World was drawn in 1587 by John White who was commander of Sir Walter Raleigh's third expedition to North America. One reason why you see photographs (and drawings) of this species so often is the fact that unlike normal butterflies—who rest with their wings folded up together—tiger swallowtails rest with wings flat, which makes them much easier to photograph.

Trumpeter swan

This is the largest native bird in the United States with a wingspan of up to eight feet and can weigh nearly 30 pounds. Once common in the world they were nearly hunted to extinction due to the popularity of their feathers which made perfect quill pens. The typewriter may have saved the species! Adults are nearly pure white while the young ones called "cygnets" are gray.

Mink

Mink prey on fish and other aquatic life, small mammals (rabbits especially), and birds, particularly water fowl. They never stray far from water. Wouldn't ya just love to curl up with one?

Killdeer

Killdeer get their name from the shrill, wailing *kill-deee* call they give so often. They lay eggs in a shallow depression that is scratched into the bare ground, and after eggs are laid the bird will often add rocks, sticks, bits of shell, and even trash to the nest. I must confess that I spent a good part of my childhood chasing after killdeer with my BB gun—but they never let me get close enough for a shot—always running just out of reach. At 54 years of age I finally "shot" one—but this time with a camera!

Indigo bunting

This bird is actually quite slim and trim but is all puffed up in this photo because it was rather chilly that day. From the look on his face I don't think he wanted his picture taken!

Bull elk grazing in the starlight

Elk are grazers, need lots of open pasture to survive in Arkansas, and will often grade all night.

White-tailed momma deer and fawn

Sometimes it takes long hours of waiting in a secluded blind, or many trips to the same location day after day, or miles and miles of hiking, and often a combination of all, before a particular wildlife viewing opportunity presents itself. And then there are times like when I took this photograph of the deer and fawn—I was driving to the store and took the picture while sitting in the front seat of my car!

I was in "wildlife" mode and had my camera with special big lens attached and all ready to shoot on the seat next to me, just in case. As I came around a corner I saw the pair standing at the edge of the woods—I slammed on the brakes, grabbed the camera, and took the picture right out the window—total time was about ten seconds. After that first photograph the deer looked up at me, then turned and bounded off into the woods. Ninety percent of the time I would have missed that shot, but I was lucky, and luck plays a major role in wildlife photography—of course, you often make your own luck!

Wild turkey eggs (previous pages)

The hen will lay a single egg each day until she is done, then sit on the nest for a full month before they all hatch within 24 hours of each other. It is amazing to me that such a large bird can sit on fragile eggs and not break any!

Red-eared slider turtle

The "slider' part of their name comes from their ability to slide off rocks and logs and into the water very quickly—you may see lots of these turtles during a float down a typical Arkansas river. This species has long been part of our domestic culture too—they are the "dime store" turtles that have been sold in pet stores for a long time. In fact this specific turtle was indeed a pet of ours, purchased when she was about three inches across. Crystal outgrew her tank at our cabin several years later and she now swims at the Elk Information Center in Jasper—stop in and say hi!

Egret watches over the nest at sunset

This photograph was taken at the edge of a large colony of more than 1,000 egret nests, using a long telephoto lens and deep red filter.

Dark-sided salamander

As a frequent cave dweller myself I've seen many types of salamanders from albino to bright orange to coal black with spots. This is the first green one I'd ever seen. It was just inside the entrance to Mystic Caverns along Scenic Hwy. 7. It is a normal color phase for this species.

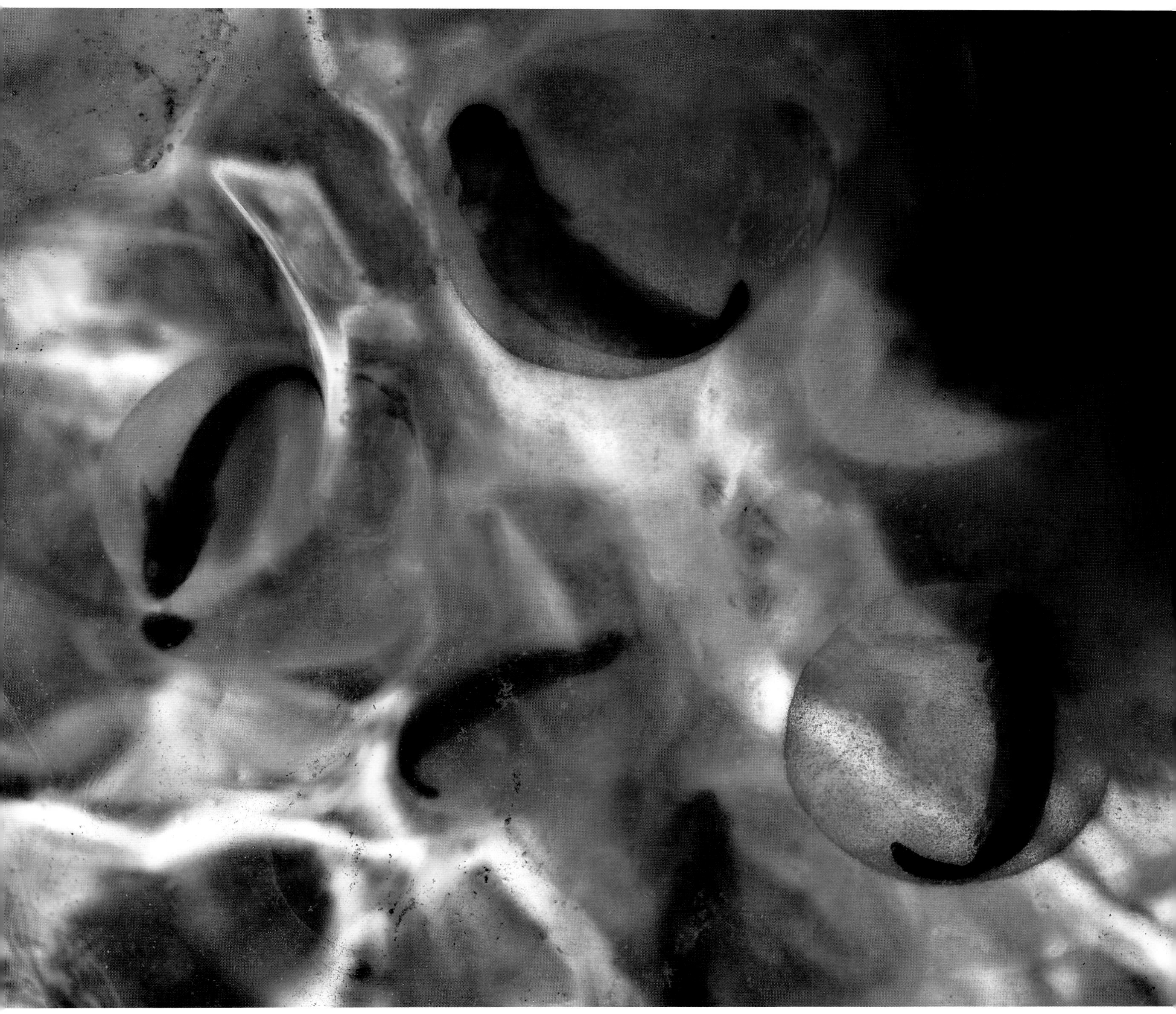

Salamander eggs

Most folks have probably seen a group of frog eggs in a mud puddle—YUK! Turns out that salamander eggs appear the same until you get down on your hands and knees and take a closer look—I wonder how many piles of frog eggs are actually salamander eggs?

Canada goose and goslings

Goslings begin communicating with mom while they are still inside the egg, and then remain with her until they are a full year old.

Bobcat

Bobcats prowl the night, and travel miles along their habitual route in search of food—you almost never see one in daylight. Like all cats, the bobcat "directly registers," meaning its hind paw prints usually fall exactly on top of its front paw prints. They love baby geese...

Woodchucks—momma with young

The average woodchuck (or groundhog—same animal) excavates hundreds of pounds of dirt while digging the den. That den may contain more than one main entrance, plus several side entrances, or "bolt holes," which are used when a quick escape from the surface is needed. There is a sleeping chamber inside the den that is lined with soft grasses, and also—a bathroom chamber!

Black bear

Bears will eat almost anything and are especially fond of ants and grubs. I've watched them roam through the forest turning over rocks and rotten logs, then licking up all the little critters they find underneath. I've seen them dig up the honeycomb of an underground bee nest (they *love* sweets!). And I once spent 30 minutes watching a bear munch on wild hickory nuts no more than 100 feet from me. I live and work in the middle of bear country and see them often—I wonder how many times they've been watching *me?*

Eastern bluebird

The sight of a bluebird always makes me smile! They are year round residents in Arkansas and can produce young three or four times a year. It is funny to sit and watch their nesting cycle. Often the male will attract a female by bringing building materials to a possible nest site, showing off by waving the material back and forth like he is some sort of great nest builder. Once he gets his girl, his job at the nest is done—the female does all of the actual nest building! But while she is sitting on the eggs, the male is in constant search of food to bring her. He will sit on a perch overlooking an open area for several minutes, then all of a sudden fly to the ground and return to the perch with a worm or grub, continue onto the nest with his prize, then return to the perch to start over again.

American toad showing off his vocal sac

Frogs are wet and slimy, toads are dry with rough "warty" skin. Toads have a beautiful song that they use to attract a mate, which is what this guy is doing. To get this photograph I laid flat on my belly in wet grass and mud, trying to focus in the dark, and waiting for him to fill his vocal sac and sing. He was shy at first but eventually I was ignored in favor of the hunt for a girlfriend. Toads have few natural predators—one reason is that they secrete a poison that ends up inside the mouths of any would-be predator that inflames the mouth and throat, causes nausea and foaming at the mouth, and most of the time the toad is spit out and that predator never touches another toad! Just ask my dog.

Bull elk in summer velvet

Members of the deer family—which include elk—shed their antlers each year and grow new ones. The new antlers are soft and tender, and covered with a skin that grows as the antlers grow. The skin has short fine hairs that look like, well "velvet." The velvet contains a network of blood vessels that nourish the growing bone underneath. Eventually the antler will quit growing and the skin dries up and is scraped off by the animal, leaving the shiny hard antler that is ready to do battle in the fall.

American kestrel

AKA "sparrow hawk," this guy is a true falcon and loves to munch on house sparrows. Unlike other falcons though, he will typically catch his prey on the ground instead of intercepting in the air. You'll see them frequently resting on powerlines in open country. They will also hover in mid air, beating their wings rapidly to stay aloft, waiting to pounce on something tasty below.

Barn owl

Thousands of wild critters are injured in Arkansas each year and could not survive without help. Thankfully there are a few dedicated individuals who are licensed by the state and medically trained to care for these animals.

Tito, the owl pictured here, owes her life to one such wildlife rehabilitater who lives near Russellville. Normally the animal is cared for until it can be released back into the wild, but Tito's injuries were such that she could never survive on her own again. Many critters like Tito become ambassadors and are used to help educate school kids and the general public about wildlife. If you ever get a chance to see a presentation at a park or school it is time well spent.

The barn owl is the only animal I know of that has a heart-shaped face—how could you not love her! And unlike some other owls, the eyes of barn owls are almost completely black, adding to their mystique. Their hearing is amazing, and in fact they can hear so well that they frequently hunt by sound, and are able to fly and catch prey in total darkness.

The spots of a three-day old white-tailed deer fawn (previous pages)

When viewed up close the spots we see on a fawn are actually clumps of individual white hairs that grow next to each other. This creates a broken pattern on the coat that helps to hide the fawn from predators by blending into tall grass or forest floor.

Lightning bugs

It would not be summer without fireflies! They are actually winged beetles, and light up for short bursts in order to attract a mate or prey. The light is produced by a type of chemical reaction called bioluminescence and is emitted from the lower abdomen. The color of the light produced in the bugs I've photographed is mostly yellow with a hint of green, and has a red band around the outside. The photograph above is a 60-second exposure at the peak of activity and shows many bugs flying through the scene.

Gray tree frog

I know he looks *green* and not gray, but gray tree frogs are highly variable in color and can change from gray to green to camouflage themselves to blend in to what they happen to be sitting on. Look close and you can see his really sticky toes which allow him to cling to just about anything.

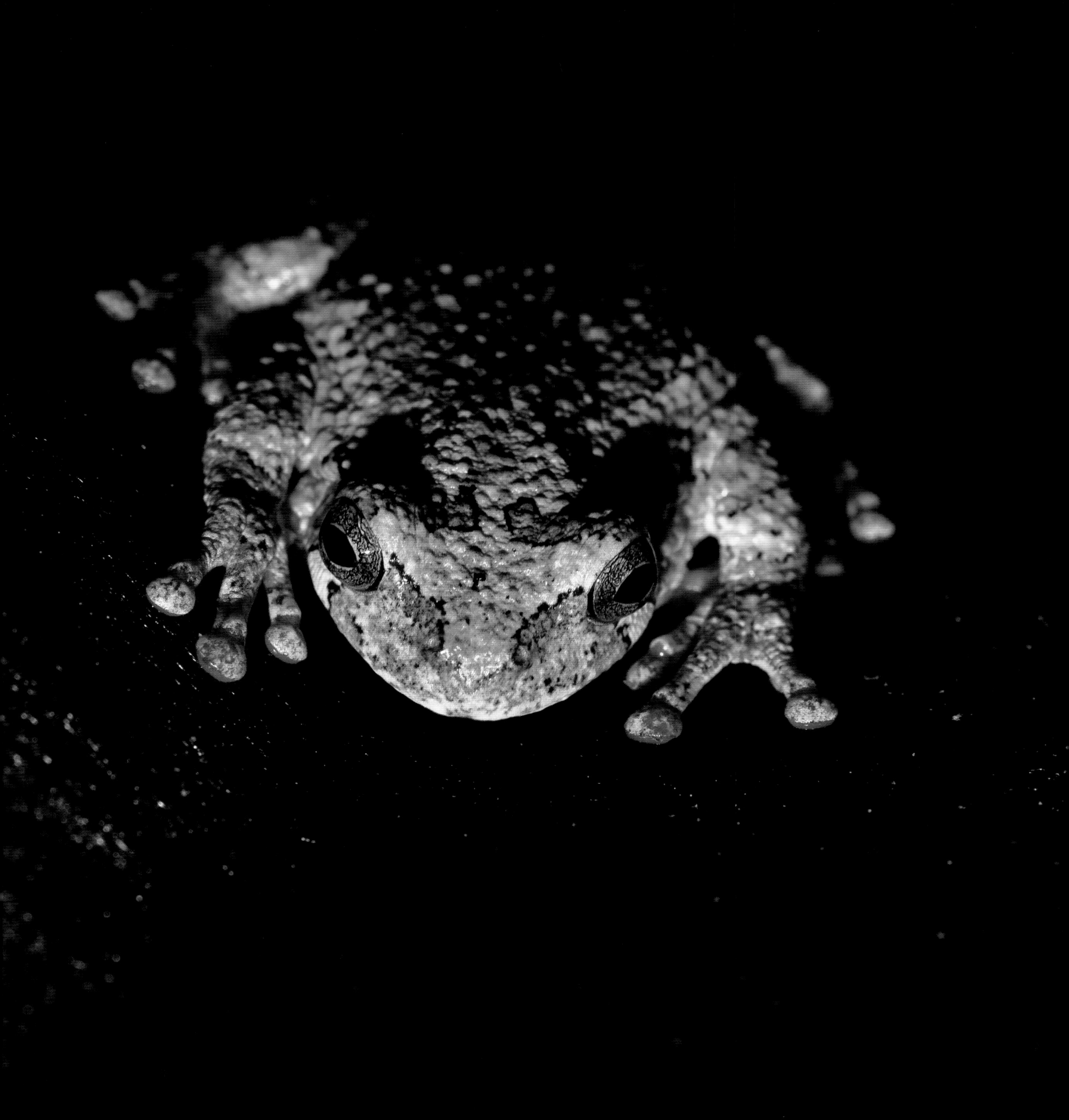

Cecropia moth

This is the largest moth species in the entire United States with a wingspread of more than six inches. I'd never seen them until one night just a few weeks before this book went to press—I found this guy clinging to the front door of my studio just waiting to be photographed. I obliged, then ran off to grab Lori Spencer's ***Arkansas Butterflies and Moths*** identification book to see what the heck he was—this guy was HUGE! When I returned to the front door the moth was gone and I never saw him again.

Polyphemus moth caterpillar
This caterpillar can eat 86,000 times its weight in less than two months.

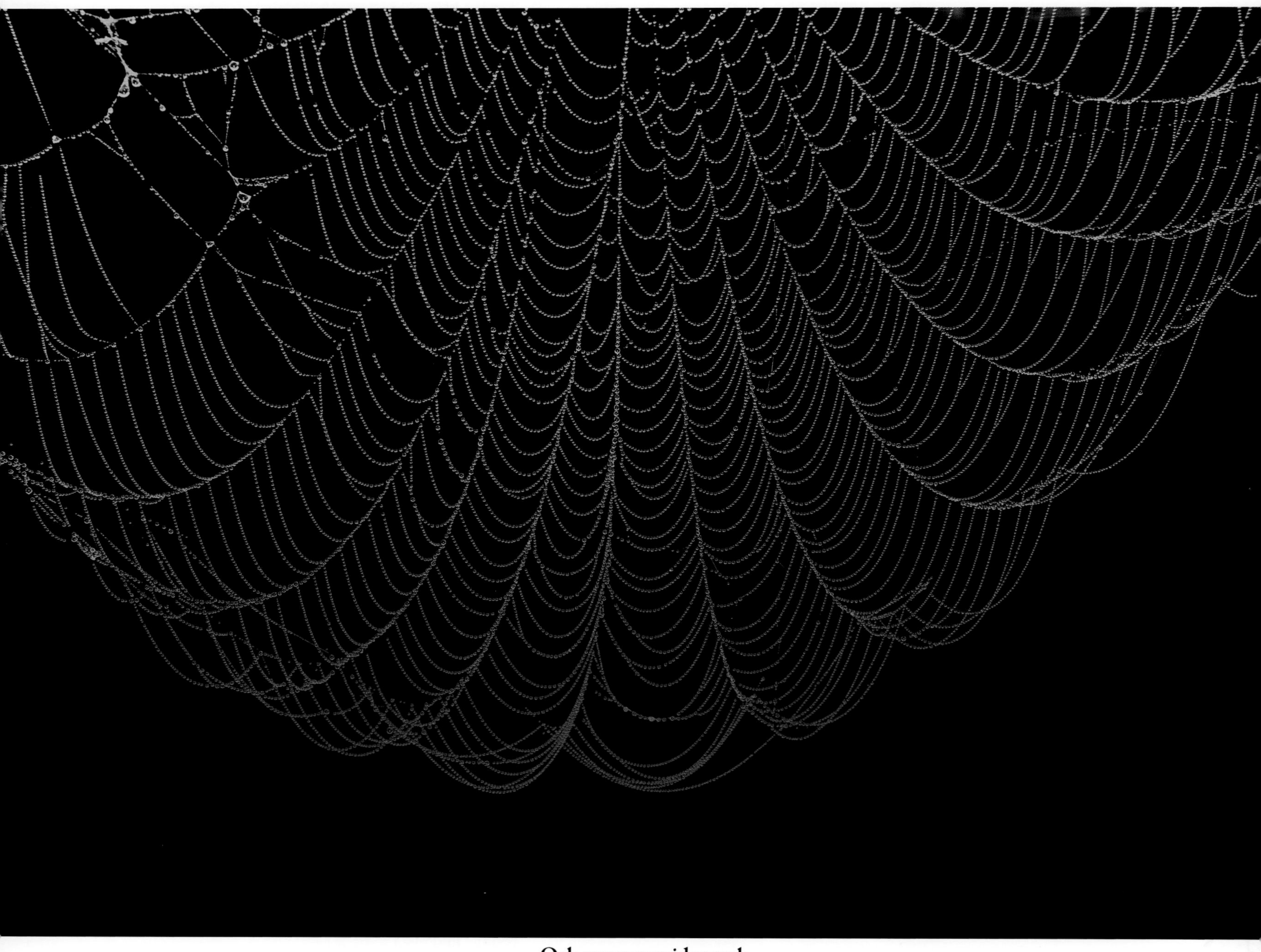

Orb weaver spider web

Many orb weavers build a new web each day. They will consume the old web in the evening, then spin a new web in the same general location—their webs are generally free of trash accumulation which is common to other spider species. The web above had collected heavy dew that formed overnight, and reflected the color of the dawn.

Snow geese and crescent moon

By flying in a "V" formation large birds like these geese can take advantage of the draft created by the bird in front.

White-tail deer buck

Deer hunting is one of the most popular sports in Arkansas. The harvesting of more than 175,000 deer annually helps to control the size of the herd so that it does not outgrow the available habitat. White-tail deer go by the names of buck (male), doe (female), and fawn. Larger members of the deer family like elk are called bull, cow, and calf. Species in other countries may be called stag and hand.

Trumpeter swan

The fading light at the end of the day often produces a golden glow that seems to mate perfectly with the beauty and grace of a swan.

Woolly apple aphids

I found this group of odd-looking critters covering up branches of an old beech tree deep in the wilderness. I was surprised to discover them associated with apple trees—then remembered there had been a large apple orchard in the area nearly 75 years before. They take their name from the woolly appearance—the individual bugs are covered with long strands of white wax that help to protect the colony of purple aphids from predators (I see yellow inside that ball of wax!).

Blue grosbeak

Grosbeaks sometimes use snakeskin as nesting material, which is thought to thwart predators. Their thick beaks are geared towards breaking up tough nuts and seeds. A group of grosbeaks are collectively known as—can you guess—a "gross" of grosbeaks.

Purple finch in the deep freeze

The purple finch is the bird that noted naturalist Roger Tory Peterson famously described as a "sparrow dipped in raspberry juice."

Ebony jewelwing damselfly

Damselflies are closely related to dragonflies and they look very much alike. The easiest way to tell them apart is to look at the wings. Dragonfly wings stick straight out from the body when the dragonfly is resting. Damselfly wings usually fold back above the body. Either species can fly up to 35 mph.

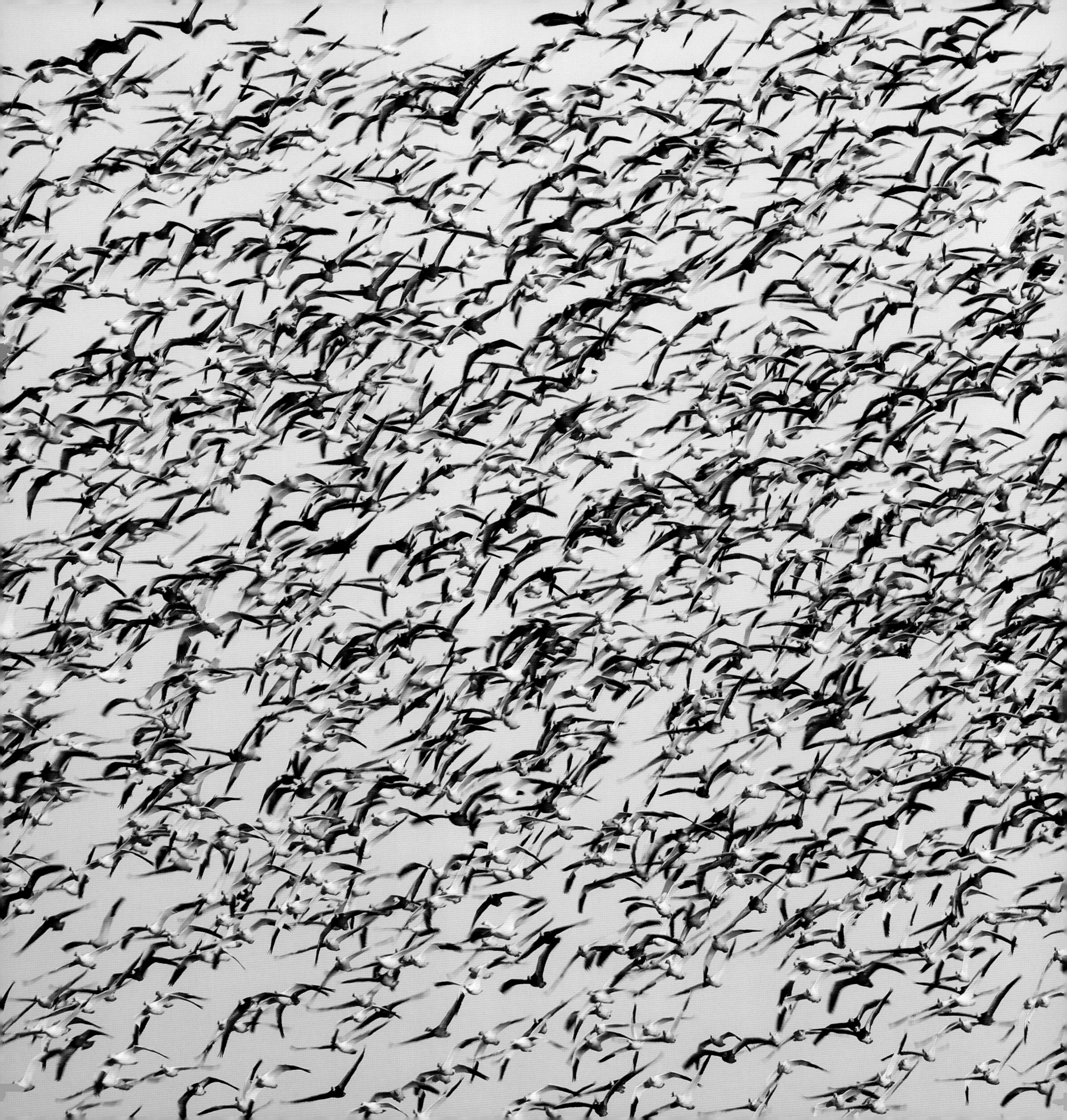

Red milk snake

I love snakes, and spent much of my childhood running wild and free in the woods near our home turning over rocks and rotten logs looking for snakes. That time in the woods taught me a great deal about the wilderness, and about myself, and I am grateful for it.

Most folks don't like snakes and have an unfounded fear of them. I'll nod my hat to ya and say that I don't like poisonous snakes at all and give them a wide berth when we meet. But far and away most snakes are kind and gentle and very helpful, and in fact many of them *eat* poisonous snakes—how could you not love them?

After some 50 years now of being in the woods I've only been bitten by a poisonous snake once—a copperhead bit my toe while I was standing on the back deck of my cabin!

One snake fact worth noting: snakes have no eyelids and instead have a transparent covering that rests over their eyes to protect their eyes from dust and dirt—this gives them a "glassy-eyed" blank appearance, which adds to their bad reputation.

I named this red milk snake Scarlet and we found her living next to our cabin. This is one of the most colorful types of snakes in Arkansas, and rarely seen since they generally are only out at night. I'm sure a lot of them are killed since they resemble one of the most dangerous snakes in the country, the coral snake—it also has a pattern that includes black and red, but it has yellow bands that touch the red bands. To tell the two species apart folks have made up several jingles, including:

Red and yellow, kill a fellow. Red and black, friend of Jack.
Red next to black, is a friend of Jack. Red next to yellow, one dead fellow.
Red touches yellow, not a nice fellow. Red touches black, bite 'em back.

You get the idea—stay away from a snake where red touches yellow!

A herd of snow geese—or is that a *flock?* (previous pages)

Can you count how many geese are in this picture? Geese migrate from far up north to spend the winter in Arkansas and often gather in flocks that number tens of thousands of birds. The darker birds in this group are snow geese too, in their dark phase—they are sometimes called a "blue goose." When a flock this size flies overhead you better run for cover! (or at least grab an umbrella)

Black bear track

A typical adult black bear in Arkansas will weigh 200 pounds or more. Some of the largest males can top 700 pounds or more. Just look at those *claws!*

Black bear, cinnamon variant

Arkansas was once known as "The Bear State" because there were so many bears. Decades of unregulated hunting for sport, subsistence, and for the meat market took its toll and the bear population was almost completely wiped out in the early 1900's. In the 1960's the Game and Fish Commission undertook a secret restocking program and brought bears from Minnesota to Arkansas that were released in at least three remote areas (White Rock Mountain and Horn Mountain areas in the Ozark National Forest, and the Dry Creek area in the Ouachita National Forest). Since that time the black bear population has been on the rise, and there are now an estimated 5,000 or more bears roaming the state. Many of the bears we see are "cinnamon" bears like this one, but most are black, and in fact some of the fur is so black and shiny that it appears dark blue.

Skipper butterfly

There are more than 3,500 species of skippers worldwide and they are very difficult to identify individual species. They get their name from the fact they skip around from plant to plant while feeding (making them difficult to photograph!). They differ from other butterfly species by their antennae tips that are hooked backward like a crochet needle instead of being blunt like a club.

Common loon

What, *loons* in Arkansas? You bet! In fact while this beautiful bird is indeed the legend of the north country, hundreds of them spend the winter floating around on the big lakes in Arkansas—they've been reported from more than 20 lakes statewide. The common loon swims underwater to catch fish (as deep as 200 feet), propelling itself with its feet. It swallows most of its prey underwater. The loon has sharp projections on the roof of its mouth and tongue that point backwards to help it keep a firm hold on slippery fish.

American alligator

Alligators have the strongest bite of any wild animal in Arkansas. Once rare here, they can now be spotted in many locations, especially in the swamps of the south. I've seen them in at least four different state parks, two wildlife management areas, and circling my little rubber boat in a small private lake. When I got out of the boat and started to wade in waist-deep water to photograph an egret nest, there were four alligators nearby keeping a close eye on me—the things I do for a photograph!

Red-tailed hawk

The scream of a red-tailed hawk is so pure and beautiful that its cry is often inaccurately used as the voice of an eagle or other hawk in Hollywood movies, or as the background for adventure scenes to give a sense of *wilderness*—and it indeed does.

Luna moth on a showy orchis

A large and beautiful moth that has such a sad life story! They have no mouth or digestive track and cannot eat. Their only job is to attract a mate and reproduce. Once the eggs are laid, they die—a week after emerging from the cocoon.

Red fox

A fox hears low-frequency sounds very well, and will locate prey by listening for their underground digging sounds, then dig them up for lunch! Their stomachs are small and fill quickly though, and a fox will keep hunting when he is full and stash extra food around his territory to feed on later.

Blue-gray gnatcatcher

By flicking its white-edged tail from side to side while going through thick brush, the gnatcatcher often scares up hiding insects which become his afternoon snack. This is one of the smallest birds we have in Arkansas and it takes a good pair of binoculars to study them up close. If you spend some time with this little bird you will have had a great day!

American bald eagle

Eagles were once extremely rare in Arkansas—in fact I never saw one until I was almost 40 years old. But now they are common in the winter months along our large rivers and lakes, and can be seen soaring high in the sky in many other rural areas. We also have dozens of nesting pairs that reside here year round, so the future looks bright for them. They love to fish, but if the truth be known, they are also a scavenger bird and can be seen on road kill and especially around some of the larger poultry operations (i.e., dead chickens). No matter where you see one, these are incredible birds and will take your breath away.

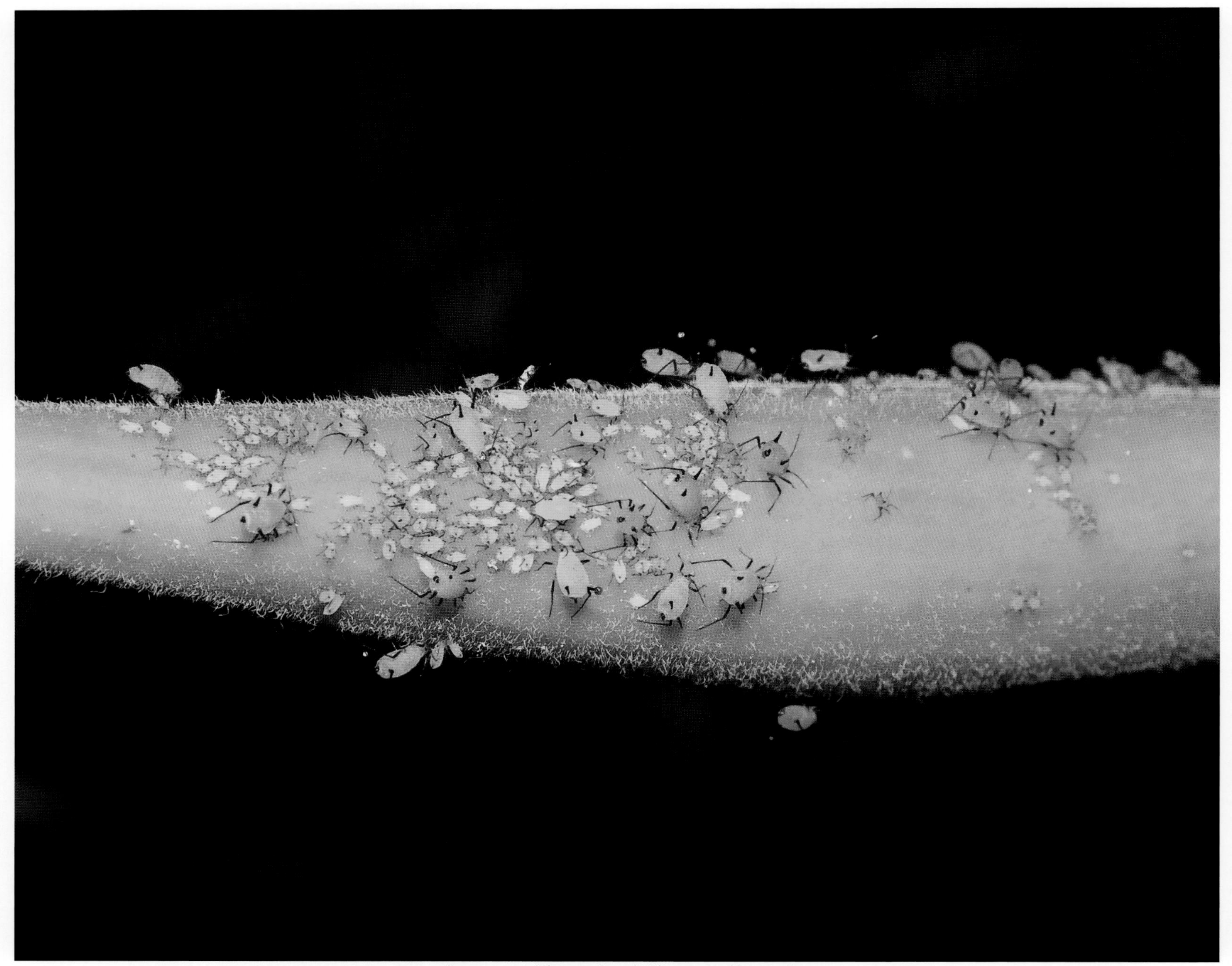

Orange milkweed aphids on milkweed

These tiny aphids are just one of more than 400 different species that are known to feed on milkweed (the monarch butterfly and its caterpillar being the most famous).

Phoebe chicks

Momma birds know not to fill the nest with eggs because eventually those eggs will become squirming chicks that will double in size every few days—they need a lot of room to grow, otherwise they will fall out of the nest! A phoebe is a small bird that builds nests under bluff overhangs so there is always a roof overhead. They look and sound a lot like the similar species, a pewee, and the way to tell them apart is that the phoebe bobs its tail while resting on a branch.

Bull elk in summer velvet at sunset

Elk used to be native to Arkansas but were driven out and hunted to extinction by the pioneers. In 1981 a restocking program began and 112 Rocky Mountain elk from Colorado and Nebraska were released in Newton County. Since then the herd has expanded into more than 500 animals that have been reported in 14 different counties. It has been a model restocking program with cooperation from private individuals, the Arkansas Game and Fish Commission, and the Rocky Mountain Elk Foundation.

One of the goals of this program was to have a sustainable elk herd large enough to support a limited hunting season, which has now been going on for many years and has produced some incredible trophy bulls.

Along the way the elk have become a tourist attraction industry in themselves. There is a large festival in Jasper each summer complete with elk-calling, dutch oven, and beauty contests. Two elk information centers in Jasper and Ponca are open year round that provide elk and other wildlife displays, area information, and guidebooks. And oh my goodness you should see Boxley Valley on a fall weekend—the elk bugle and fight just like they do in Yellowstone, and the traffic jams and tailgate party atmosphere is like Yellowstone too!

What a thrill it is to hear an elk bugle that echoes through the valley on a crisp fall evening—it is one of the greatest wildlife experiences in the state. Elk viewing areas include Boxley Valley, Erbie, and the lower Richland Creek area, all along the Buffalo River. Elk are grazing animals, and so spotting them out in pastures *very early* or late in the day is easy. Stop by an elk center to find out the best viewing locations and times.

The photograph of the big bull at right was taken in early summer while the antlers were in the "velvet" stage of rapid growth. The big boys separate from the herd in the spring and head for the high ridgetop pastures, then return to the valleys in the fall to assemble their harem, which can contain many dozens of cows. A big bull like this one can weigh 1,000 pounds and is the largest wildlife species in Arkansas.

Photo tip—to make a silhouette when you have a very bright background, exposure for the background and the foreground will naturally silhouette.

Barred owl (previous pages)

The barred owl is a highly vocal owl giving a loud and resounding "hoo, hoo, too-HOO; hoo, hoo, too-HOO, ooo" which kind of sounds like "Who, cooks, for-you? Who, cooks, for-you'alllllllll?" I have spent countless evenings on the back porch listening to their music (they will sometimes answer if you call out to them), and have met up with them many times in the woods—this is one of our most common deep-woods owls in Arkansas.

Baltimore oriole, male

To attract these guys to your feeder use grape jelly or orange slices—this pair ate an entire bottle of jelly in one day at our house!

Baltimore oriole, female
The Orioles have won the World Series three times.

American robin eggs

The first recorded use of "robin egg blue" as a color name in English was in 1873. It has been long trademarked by Tiffany & Co., and is an official crayola color.

Black bear

The term "lumbering" must have been coined to describe how an old bear moves along, taking his sweet time.

White variant of a gray squirrel

In all my years in the woods I had never seen a white squirrel before! I thought they were a freak of nature and confined to a couple of urban areas elsewhere. But it turns out we have several populations of white squirrels in Arkansas, and dozens of reports of white individuals as well. We started seeing a single white squirrel in January this past year near our cabin, and continued to see him several times a week through May when he disappeared. Both white and black squirrels are simple variants of a normal gray squirrel, although some are true albinos and have pink eyes.

Osprey being chased by the moon

An osprey is the only raptor whose outer toe is reversible, allowing it to grasp its prey with two toes in front and two behind. It sometimes goes underwater in search of fish, and can close its nostrils to keep out water during dives, and backwards-facing scales on the talons which act as barbs to help hold its catch. In flight, the Osprey has arched wings and drooping "hands," giving it the gull-like appearance you see here. I used a very long telephoto lens (equivalent to 1500mm) and then cropped the resulting image—that is why the moon appears so large.

Indigo bunting

Indigo buntings migrate at night, using the stars for guidance. They learn orientation from experience as young birds observing the stars. Even though they are one of the most BLUE things you will ever see in the wild, they have no blue pigment—they are actually black, but the diffraction of light through the structure of the feathers makes them appear blue.

Coyote

Coyotes can run 40 miles an hour. They will catch and eat voles, rabbits, chipmunks, mice, birds, snakes, lizards, large insects, and even small deer. In other words, they eat as much as a human teenager! I've spent many winter nights sitting next to a campfire listening to the lonely howl of a distant coyote, and can relate to their solitary wandering nature.

Eastern meadowlark

Meadowlark nests are constructed by the female, and are built on the ground of grasses woven into surrounding vegetation. Sometimes the nest can have an arch or roof, and even a runway leading to the opening. The male will typically have two mates at the same time. They are not actually "larks" but rather in the same family as blackbirds and orioles (Alaudidae).

White-tailed deer bucks during the rut

During the breeding season or "rut" a mature buck will create a "scrape" line bordering his territory—he will paw and scrape down to bare dirt in several places (two-three feet in diameter), tear up the ground, and urinate in the middle—this is his calling card to let any available doe know he is in the area. He will run from scrape to scrape around the clock, checking to see if any does have left their scent—if so, it's party time!

Leather-wing feasting on a passion flower

If you look really close at one of these bugs the wings really do look just like leather! They seem to be drawn to beautiful wildflowers—I have photographed them on many different species of bright and colorful wildflowers.

Eastern screech owl

An owl cannot move its eyes—it must turn its head to watch a moving object. The normal territorial call of a screech owl is not a hoot as with most typical owls, but a trill consisting of more than four individual calls per second given in rapid succession. When courting, the pair has a special song they sing together as a duet. This little owl got a bum wrap in the movie "My Cousin Vinny"—the loud and obnoxious scream was definitely not made by a screech owl as was shown. Their voice is much softer and pleasant.

Cougar—puma—panther—catamount—mountain lion

Dr. Neil Compton once told me about a large pile of scat he collected near Hawksbill Crag in the 1970's that he sent off to an expert for identification—it was confirmed to be cougar scat. Cougars have always been rare in Arkansas, and our Game and Fish Commission holds that we've never had a "wild breeding population" in modern times.

Do we have them here now? That depends on who you ask. I've heard dozens and dozens of reports from friends and colleagues who I respect that have independently seen a cougar in the wild in Arkansas. But for me personally I always found it hard to believe that with as much time as I spend in the deep woods each year—at all times of the day and night—that I never saw one or even any sign of one. That was until just a few months ago—I got a clear look at a cougar in the wild, and it just happened to be within a quarter mile of where Dr. Compton collected that confirmed cougar scat. None of these sightings prove anything—other than we do seem to have cougars in the wild in Arkansas, at least one of them, and probably more.

There has been talk lately of an attempt to relocate some cougars to Arkansas from a wild population in Florida whose habitat is being destroyed beyond its ability to support the big cats. Arkansas is a perfect location for them due to the rugged and large remote tracts of wilderness, and the fact that we have a good supply of their favorite food—white-tailed deer. Many folks are opposed to this for a variety of reasons and who knows if it will ever happen. I'm not shy about telling my personal opinion on this—I sit squarely on the fence! We shall see.

A photograph would not prove anything one way or another about a "wild, breeding population" in Arkansas. One cat does not a population make. For the record, I am not claiming that this photo is of a wild cougar in Arkansas—it is not.

American bald eagle comes in for a landing (previous pages)

I know that Ben Franklin wanted the turkey, but our national's symbol truly is a majestic and incredible wild figure and I never tire of seeing them. On this particular day I sat in a blind for more than eight hours to photograph eagles in flight. During all that time there was a period of only 30 minutes of eagle activity and I made thousands of exposures using a long telephoto lens and a camera that could shoot many frames a second. I spent the rest of the time anxiously waiting for more (never happened—not even another single picture), and wondering if I had any good photographs.

WILDLIFE PHOTOGRAPHY

Being a wildlife photographer is easy. First, you need to lift weights—the equipment is heavy. Second, you need a second job—the equipment is expensive. Third, you must have an understanding spouse, significant other, children, parents, and boss—it is time consuming. Fourth, you must have thick skin and thrive on failure because 99% of all your images will be garbage. And fifth, you must be really lucky—timing is everything, and with wildlife photography a split second can mean the difference between a work of art and a snapshot. Of course, you make your own luck.

I've always admired wildlife photographers for their dedication, and because they are able to produce some of the most amazing images you'll ever see. What I discovered when I decided to become a wildlife photographer was that one of the most incredible things is that you get to spend so much time with the wildlife, on their turf, become a part of their lives, and they become a part of your life. It is a remarkable situation quite unlike any other I've encountered in more than 30 years as a nature photographer. It is addictive.

The equipment. It is heavy and expensive and can be difficult to work with, and the brand is not important. Most of the time you will need more expensive and complicated equipment than you would need to photograph a waterfall or a rock. For small birds and big game that is shy and only active in dim light, you will need long and really fast glass, and that means big bucks and lots of weight. I used both Nikon and Canon 300, 400, 500, and 600mm lenses with 1.4x and 1.7x teleconverters on their highest-resolution bodies for much of the work in this book. A very high shutter speed of 1/1000th of a second or faster was often required to stop movement—for that, and for shooting in very low light you need a camera that can shoot at really high ISO (up to 3200 for the elk in starlight image). I also made many of the images in this book with a 5 megapixel Minolta point and shoot camera, but some with a 39 megapixel Phase One system as well. And for the bugs, a dedicated macro system will produce the best images, although you probably don't need to be the macro junkie that I am—I have no less than seven macro lenses.

I used a tripod for most everything except for the point and shoots. I never used filters, other than the image of the egret at sunset—used a deep red filter for that one. Flash was rarely used—only for some of the macros. I used a camouflage blind much of the time—you sometimes have to spend a lot of time in the field before an animal will get close to you, and getting close is important. A lot of the images were made while sitting in the car too—I always had a camera and big lens on the front seat ready to shoot. Some of the images in this book were created within a ten-second window of opportunity—if you are not prepared, there is no image.

Every single image in this book has been optimized in Photoshop. I want to produce final images that are as close to what I envisioned as possible, and no camera can do that without the human touch of further digital processing (examples: I blended 11 images together that were focused on different parts of the flower to get page 134 sharp—no single image could get everything in focus; warm evening light turned the swan on page 99 yellow and I had to manually return the swan to natural white while maintaining color in the reflection). Careful processing is a must!

Successful wildlife photography is a combination of being there at the right time with good light, getting close, having the right equipment and knowing how to use it, and being able to do the processing (plus lots of luck!). To master your equipment, become proficient in creating great photographs, and learn an efficient digital darkroom workflow, come to one of my workshops or purchase a DVD tutorial. To see what is available visit www.Cloudland.net. And be sure to have your camera ready!

Clockwise from upper left: osprey and largemouth bass, northern cardinal, Indigo bunting, prothonotary warbler

THE PHOTOGRAPHER

Tim Ernst, 54, lives in a log cabin called Cloudland in the middle of the Buffalo River Wilderness in Newton County, Arkansas, with wife, Pam, and daughter, Amber.

I've been a professional nature photographer for more than 34 years with images published in most of the major nature publications from ***National Geographic*** on down, including hundreds of national, regional, and local magazines, books, and calendars. This is my tenth coffee table picture book, and the first to be filled with all wildlife photography. I have written a couple dozen guidebooks to outdoor Arkansas destinations that will lead you to waterfalls, hiking trails, and special scenic locations. My ***Arkansas Nature Lover's Guidebook*** includes many great wildlife-viewing spots (where a lot of the images in this picture book were made). My lovely bride and I own and operate a small publishing business, **Cloudland.net Publishing**, now in its 28th year of operation. I also sell fine art prints on traditional photographic paper or on incredible gallery-wrapped canvas to businesses and individuals around the country via our online galleries, and through our Buffalo River Gallery location that serves as gallery, digital darkroom, and print studio. And I've been teaching nature photography workshops to photographers of all skill levels for the past 24 years, and now offer DVD tutorials as well.

To see or order any of our products, view a schedule of my unique slide programs that I give around the region, get more information about my workshops and tutorial DVDs, or to keep up with life in the wilderness via our ***Cloudland Cabin Journal***, visit us on the web at www.Cloudland.net, or for special online galleries of my photography visit www.BuffaloRiverGallery.com.

Other books by Tim Ernst
Arkansas Portfolio picture book
Wilderness Reflections picture book
Buffalo River Wilderness picture book
Arkansas Spring picture book
Arkansas Wilderness picture book
Arkansas Portfolio II picture book
Buffalo River Dreams picture book
Arkansas Waterfalls picture book
Arkansas Landscapes picture book
Arkansas Nature Lover's guidebook
Arkansas Hiking Trails guidebook
Arkansas Waterfalls guidebook
Ozark Highlands Trail guidebook
Buffalo River Hiking Trails guidebook
Ouachita Trail guidebook
Arkansas Dayhikes guidebook
The Search For Haley
The Cloudland Journal

Cattle egret in breeding plumage

While I would probably choose a swift red-tailed hawk or majestic bald eagle to be if I were a bird, the reality of life is that I can most often relate to this fellow on the right—a typical hair day for me, and my office usually looks just like this nest!

Bull elk in summer velvet—sometimes ya just gotta scratch that itch!